W9-ANF-085

3 1486 00331 3347

12/15

ROBOT WORLD
ROBOTS AT HOME

by Jenny Fretland VanVoorst

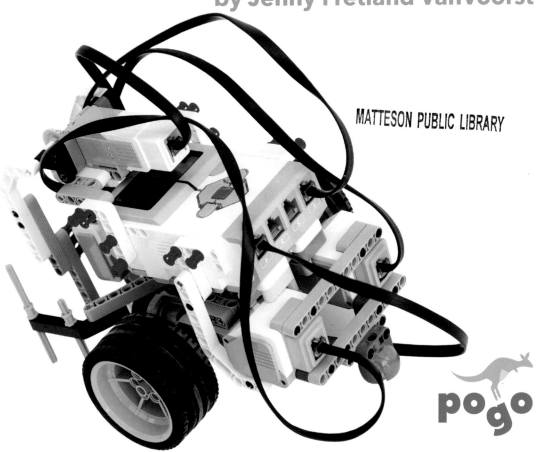

MATTESON PUBLIC LIBRARY

pogo

Ideas for Parents and Teachers

Pogo Books let children practice reading informational text while introducing them to nonfiction features such as headings, labels, sidebars, maps, and diagrams, as well as a table of contents, glossary, and index.

Carefully leveled text with a strong photo match offers early fluent readers the support they need to succeed.

Before Reading

- "Walk" through the book and point out the various nonfiction features. Ask the student what purpose each feature serves.
- Look at the glossary together. Read and discuss the words.

Read the Book

- Have the child read the book independently.
- Invite him or her to list questions that arise from reading.

After Reading

- Discuss the child's questions. Talk about how he or she might find answers to those questions.
- Prompt the child to think more. Ask: Do you know anyone who has a robot in his or her home? If you could have a robot to do one daily task for you, what would it be?

Pogo Books are published by Jump!
5357 Penn Avenue South
Minneapolis, MN 55419
www.jumplibrary.com

Copyright © 2016 Jump!
International copyright reserved in all countries. No part of this book may be reproduced in any form without written permission from the publisher.

Library of Congress Cataloging-in-Publication Data

Fretland VanVoorst, Jenny, 1972-
   Robots at home / by Jenny Fretland VanVoorst.
      pages cm. – (Robot world)
   Includes bibliographical references and index.
   ISBN 978-1-62031-216-2 (hardcover: alk. paper) –
   ISBN 978-1-62496-303-2 (ebook)
   1. Robots–Juvenile literature. I. Title.
   TJ211.2.F745 2015
   629.8'92–dc23

                                      2015020997

Series Designer: Anna Peterson
Book Designer: Michelle Sonnek
Photo Researcher: Michelle Sonnek

Photo Credits: Alamy, 8-9, 20-21; Corbis 18-19; Dreamstime, 11, 12-13; Getty, 16-17; iStock, 1, 5, 6-7, 23; Shutterstock, cover, 3, 4, 15; Thinkstock, 10.

Printed in the United States of America at Corporate Graphics in North Mankato, Minnesota.

TABLE OF CONTENTS

CHAPTER 1

WHAT DO THEY DO?

What can vacuum your floor, drive you to school, and even play a game of fetch?

A robot! Robots are machines, but they are so much more. They are machines that can sense, think, and act.

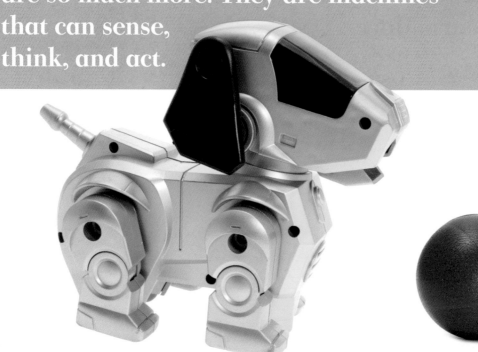

We use robots for many things. Robots who work in the home usually do jobs that people would rather not do.

pool cleaning robot

There are some chores that are just plain dull. But for a robot, there is no such thing as boring. They can do the same task over and over again. They don't get tired or sloppy. And they don't even ask for an allowance!

robotic
lawn mower

There are robots that vacuum and wash your floors. They mow lawns and clean pools.

Some robots can serve you a meal. Others protect your house when you are away.

DID YOU KNOW?

Household robots exist in a world built for people. That's why so many of them are built to look like us. Robots that look like people are called **androids**.

CHAPTER 2

HOW DO THEY WORK?

Robots follow **programs** that tell them what to do. **Sensors** help them engage with the world.

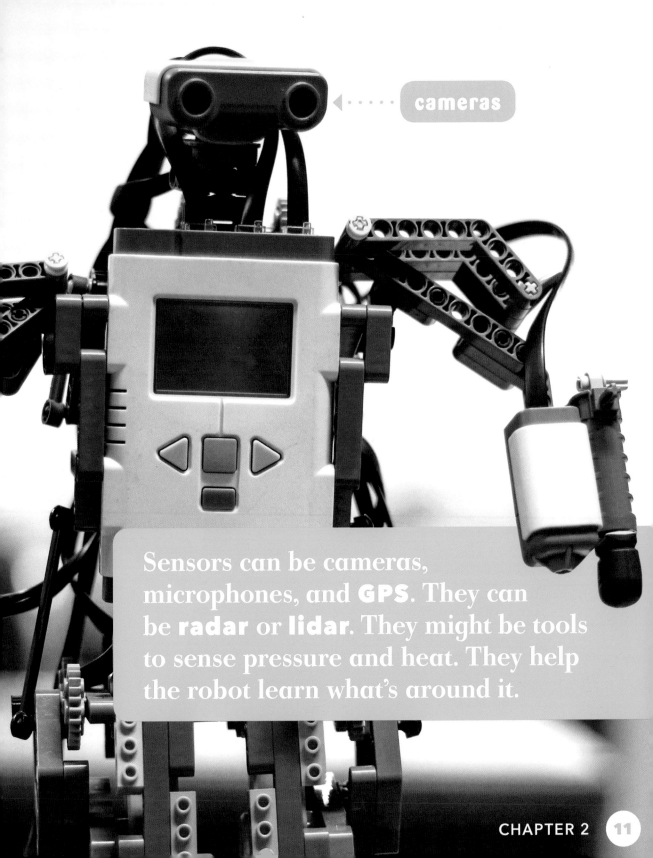

cameras

Sensors can be cameras, microphones, and **GPS**. They can be **radar** or **lidar**. They might be tools to sense pressure and heat. They help the robot learn what's around it.

A robot's computer puts together all the **data**. It makes a picture of the world. It makes a plan. And finally, the body of the robot takes action. This is called the **sense-think-act cycle**.

computer

CHAPTER 3

MEET THE ROBOTS

Robot vacuum cleaners use sensors to find their way around your house.

They move across the floor until they sense something in their way. Then they change direction. It's not a very efficient way to work. But sooner or later everything gets done.

lidar (laser sensor)

camera

Wouldn't it be cool if you had a car that could drive itself? You probably will someday! Self-driving cars sound like science fiction. But they are a lot closer to science fact.

TAKE A LOOK!

Scientists think that by 2030 there will be more robots on the planet than people! Wow!

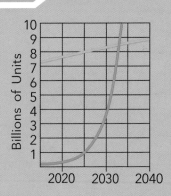

■ = robots
☐ = people

Several companies are working on self-driving cars. The cars don't even have a driver's seat. You program in where you want to go. Sensors keep the car in its lane. They keep track of distance from other cars. You just sit back and enjoy the ride!

Robots can be fun toys as well. There are robotic soccer competitions and robotic LEGO sets that you program yourself.

Do you have a robot at home? Just wait! You will soon.

ACTIVITIES & TOOLS

ROBOT DESIGNER

Think of a task that you need to do often that you dislike. It might be making your bed, setting the table, or folding laundry.

Think about how you might design a machine to do that for you. What would it need? Cameras to see? Would it move on wheels, treads, feet, or something else? Would it have an arm? One or two?

Then create a description or illustration of your robot, including:

- A name
- What materials it is made of
- What sensors it uses
- How it looks and how it moves
- What it can and can't do

You might ask a friend to create a robot for the same task. Then compare your robot designs. How are they similar or different?

GLOSSARY

android: A robot built to look and function like a human being.

data: Facts about something that can be used in calculating, reasoning, or planning.

GPS: A navigation system that uses satellite signals to find the location of a radio receiver on or above the earth's surface; abbreviation of global positioning system.

lidar: A device that uses laser beams to detect and locate objects.

program: A set of instructions that a robot follows.

radar: A device that uses radio waves to detect and locate objects.

sense-think-act cycle: The process by which a robot collects data using sensors, formulates a plan using its computer, and enacts the plan using its physical body.

sensors: Onboard tools that serve as a robot's eyes, ears, and other sense organs so that the robot can create a picture of the environment in which it operates.

INDEX

TO LEARN MORE

Learning more is as easy as 1, 2, 3.

1) Go to www.factsurfer.com

2) Enter "robotsathome" into the search box.

3) Click the "Surf" to see a list of websites.

With factsurfer, finding more information is just a click away.